Erasmus Darwin Hudson

Doctors, Hygiene and Therapeutics

An Anniversary Discourse Delivered before the New York Academy of

Medicine, November 18, 1875

Erasmus Darwin Hudson

Doctors, Hygiene and Therapeutics
An Anniversary Discourse Delivered before the New York Academy of Medicine,
November 18, 1875

ISBN/EAN: 9783337816407

Printed in Europe, USA, Canada, Australia, Japan

Cover: Foto ©berggeist007 / pixelio.de

More available books at **www.hansebooks.com**

DELIVERED BEFORE THE NEW YORK ACADEMY OF
MEDICINE, NOVEMBER 18, 1875.

BY

E. DARWIN HUDSON, Jr., A. B., M. D.,

PROFESSOR OF THE PRINCIPLES AND PRACTICE OF MEDICINE, WOMAN'S MEDICAL COLLEGE
OF THE NEW YORK INFIRMARY; FELLOW OF THE NEW YORK ACADEMY OF
MEDICINE ; MEMBER OF THE MEDICAL SOCIETY OF THE COUNTY
OF NEW YORK.

UNA FIDES ALTARE COMMUNE.

NEW YORK:
D. APPLETON AND COMPANY,
549 & 551 BROADWAY.
1876.

NEW YORK ACADEMY OF MEDICINE.

ANNIVERSARY ADDRESS.

MR. PRESIDENT AND FELLOWS OF THE ACADEMY OF MEDICINE.

I must assume that the honor conferred upon me by
this learned and honorable Academy, in electing me your
orator, is a greeting from the older members who have
founded it, and advanced it to its present position of dignity,
influence, and prosperity, to the younger generation of its Fellows—to the younger generation of physicians—an assurance
of your interest in our education, our growth in experience and
skill, our reputation and success ; that as Fellows, the older
and younger, though in age rather as fathers and sons, are but
the senior and junior members of our honored and noble
brotherhood, and that as physicians, we have a common cause,
a single interest, " *Una fides altare commune,*" the advancement in integrity and usefulness of the science of medicine.

The Academy of Medicine, instituted in 1847, to-day for
the first time celebrates its anniversary in its own hall, has a
home of its own, a building dedicated to the healing art ; not
so imposing as the Serapeum at Alexandria, or the Temples of
Æsculapius, but an edifice adorned, capacious, and inviting.
May it prove a centralizing, harmonizing, organizing power
in our profession ! The Royal College of Physicians first occu-
pied the residence of Linacre, its founder ; its now celebrated
museum and library started from the gift of the great Harvey.
This mansion is the joint contribution of the Fellows of this
Academy ; our library—a nucleus of American medical litera-

ture, unique and full—is the gift of its President. The Academy is to be congratulated on its present success. What may be its future growth! What its future influence for good, for the advancement of the profession, founded, as it is, upon wise and lofty principles! How comprehensive of the needs of the profession are the few, well-defined objects of this Academy:

1. The cultivation and advancement of the science of medicine.

2. The promotion of the character and honor of the profession.

3. The elevation of the standard of medical education.

4. Public hygiene.

The Academy imposes no limitations of constitution or creed, unduly conservative or violating the liberal and progressive tendencies of our day. It was planned in a catholic spirit; it seeks the welfare of the physician, and of the public which he guards and guides.

I would gladly present, on the occasion of this first anniversary in our new home, something worthy of the time and place. It has been the custom of the scholarly men, who in the past have filled this honorable position, to consider some broad subject, instructive to the profession and its friends, or seeking to right abuses and correct popular misapprehensions —subjects which their erudition, their age, and long experience well fitted them to discuss. Lacking these qualifications, I feel, in the words of Sidney, that "all is but lip-wisdom which wants experience."

But, " the youngest heart has the same waves in it as the oldest, though without the plummet which can measure their depths."

Our profession, as it exists to-day—its *personnel*, its ethics, its *esprit de corps;* its standing with the learned and with society, and the resources with which it practices to cure—in a word, our profession, its honest value as a healing art, its public influence; these are questions of to-day, as of the past. For now, as then, our profession is our alter ego, in which we live, and move, and have our being.

Dr. Samuel Johnson said : "A physician in a great city seems to be the mere plaything of Fortune ; his degree of reputation is, for the most part, totally casual ; they that employ him know not his excellence, they that reject him know not his deficiency." The same is true to-day. There is the same relation with the public, who can have no certain rule to guide them in discriminating the certain and scientific from the pretentious and false. Charlatanism has its friends among the learned and cultured—those who in their own fields of labor, and in the affairs of the world, are esteemed judicious and honest. Even the great Bacon was not averse to dabbling in every specious and presuming mode of cure. "The impostor frequently triumphs at the bedside of the sick when true merit is affronted and dishonored." In the middle ages, we are told by Hecker, the empirics "boasted to the inhabitants of their experience and skill, and with their pills and hellish electuaries flitted about from place to place, especially where rich merchants were to be found, from whom, should they be cured, they obtained the promise of mines of gold." In France, the charlatan is described as an affair of "*pompeux galimatias, spécieux babil, des mots pour des raisons, et des promesses pour des effets.*" "A physician," we are told by the author of "Physic and Physicians," "should never affect ignorance of the cause of a complaint. He should place it in the pancreas or pineal gland, if he has no other local habitation ready at the moment."

A quack being asked how, without knowledge, he thrived, took his interrogator to the window, and asked, how many of a hundred passers-by were wise? "Perhaps one," was the reply. "Well, the ninety-nine are mine." What wondrous cures are wrought by the charlatan and nostrum-vender in these days! What testimonials do they present from men of business, estimable ladies, and retired clergymen, scarcely less ludicrous than the caricatures by Matthews, in the "Humors of a Country Fair!" "Sir, I was cut in two in a saw-pit, and cured by one bottle ; " "Sir, by the bursting of a powder mill, I was blown into ten thousand anatomies ; the first bottle of your incomparable collected all the parts together, the second restored life and animation, before a third was finished I was in my usual state of health." We have to-day clairvoyant doctors, who, like Paracelsus, think, since God has not imparted to us the secrets of medicine, it is justifiable to consult the devil. More fortunate is pretension when it assumes an organized form, sets up a universal dogma, asserts refinement in treatment, demands reform. "*Similia similibus*"—the rust of Telephus's sword cured the injuries it inflicted. Sir Theodore Mayerne banished hypochondria with the balsam of bats. Hudibras declares—

> "Wounds by wider wounds are healed,
> And poisons by themselves expelled."

Whether it be "a specious mode of doing nothing," "the art of amusing the patient while Nature cures the disease," or the vended and heroic nostrum of the day, all find adherents. The wise and foolish alike follow—

> "Each proselyte would vote his doctor best,
> With absolute exclusion of the rest."

But the real success of charlatans is transient. They prosper for a brief period and disappear. Nostrums are popular and decline. Organized schools of theory and dogma have had a brief existence in the past; those of our day will also pass away. The aphorism of Hippocrates, that "life is short, but art is long," is prophetic of the perpetuity of our art. "Skepticism of medicine," says Mackness, "has become a sort of literary tradition." But the sneer of Montaigne, the satire of Molière and Rousseau, the derision of Dryden and Pope, the caricatures of Dickens. expose only our personal defects and foibles; the rich and poor, the learned and ignorant, look to us, with faith and hope, in their hours of trouble. And how true the words of Byron for all our adversaries:

> "Physicians mend or end us,
> Secundum artem—but though we sneer
> In health, when sick, we call them to attend us,
> Without the least propensity to jeer."

Yet what means have the public of judging between doctors of the same or different schools of practice? what test of the treatment they apply? Even success is no criterion, for the curative results may have been due to Nature rather than remedies, the diagnosis erroneous or falsely made, exaggerating the fleeting disorder into grave disease. "The purse of the patient frequently protracts his cure." There is but one protection for an ill-informed and indiscriminating people—the presence in the community of educated, scientific physicians, who have attained a high standard, prescribed and enforced by law; men who, by virtue of rigid mental discipline and thorough culture, must be believed to possess personal integrity, and loyalty to scientific truth. So long as standards are low or indefinite, and the entrance to the profession has few restrictions, the choice of a doctor will be an experiment, and human health and life will be at the mercy of inexperience, ignorance, error, and deception. But when all physicians are men of trained minds and high acquirements, as the public recognize that our standards are high, confidence will strengthen; only regularly-educated men will be demanded; theorists, empirics, boasters, self-announcing doctors, will find but a poor and shifting patronage.

The practical question for our profession to-day is the higher education of the doctor, the only means of combating quackery. We fail to rival the adherents of charlatanism in mannerism and display; we must trust to solid education of the individual, scientific study, sound judgment, superior culture, as a profession. We must be actuated by honor, and work with unity of purpose.

That too many are wanting in the confidence and equipoise which command esteem, that many have not the power of translating their higher technical knowledge into succinct and intelligible terms which are lucid and convincing, the public have frequent opportunity to see. Witness the half-weighed statements of our medico-legal experts, the conflicting testimony on insanity. From the disagreements of doctors, pitted one against the other, the intelligent layman must infer the ignorance and avarice of the man, or an unsettled, indefinite state of his art. Felicity and simplicity of expression are gifts of the learned mind ; they bear the *sigillum simplex veri*—the simple stamp of truth. But the medical profession has always had the strength of knowing its own weakness ; even Hippocrates, while praising his art, lamented the ignorance of those who practised it.

And first, the material with which the profession of the present and future is being reënforced. Huxley, an admirer of its generous tendencies and comprehensive field, has called attention to the numbers of the untrained and unscholarly who choose to study the profession of medicine. Should not the young man be a man of letters, possessed of scientific tastes and knowledge ? Should he not have the basis of broad, liberal education ? Says Dr. John Brown, the author of " Locke and Sydenham " and of " Rab and his Friends :" " I give my vote for going back to the old, manly, intellectual and literary culture of the days of Sydenham, Arbuthnot, and Gregory ; when a physician fed, enlarged, and quickened his entire nature ; when he lived in the world of letters as a freeholder, and reverenced the ancients, while at the same time he pushed on among his fellows, and lived in the present." A retrospect of medicine will convince any one that our medical men of letters have strengthened our profession in the esteem of the educated world ; at the same time they have been among the contributors to its special literature. Celsus, the " Cicero of Medicine," Erasmus Darwin, Goldsmith, Keats, Akenside, Arbuthnot, Sir James Clark, Sir Henry Halford, Abercrombie, Sir Charles Bell, Faraday—have they done nothing by their non professional writings, their beauty of diction and general discourses, to honor their profession and secure the sympathy and fellowship of men of science and letters ? It has been said that medicine is the foster-mother of natural science, and Dugald Stewart regards it the best preparation for the study of the mind. Most of the great thinkers—Socrates, Aristotle, Locke, Descartes, Sir William Hamilton—have considered medicine their favorite study. Surely, the latter believed it had the true elements of science; as defined by himself, "a complement of cognitions, having in point of form the character of logical perfection, and in point of matter the character of real truth."

Reversely, what can the physician profit by companionship with these students of the philosophy and history of matter and mind? Is not the doctor who has traced the history of the world's mutations, the lives of nations and individuals, a better judge of human character? Does not his study of mental and moral philosophy make him a better alienist, enable him to read the influence of thought and emotion upon the body in health and disease? Surely, ours is a learned profession, though it be a vocation of labor, whose mastery exceeds the possibilities of any one man's life, whose principles and truths proclaim it a science. Our physicians should, therefore, have the personal culture and scholarly attainments which would unite them to all classes of earnest and scholarly men. "The strength of all sciences which consisteth in their harmony, each supporting the other, is the strength of the old man's fagots in the band" (Bacon). Perhaps in our country we have had no one who has filled the full measure of the learned and cultivated physician so completely as Dr. Benjamin Rush. He was educated at home and in Europe. Returning to Philadelphia, he became eminent, equally as physician, citizen, and scholar. As a physician he was a professor of medicine, the founder of the first dispensary of this country, a practitioner of unequaled clientage, and an original investigator of disease. To his personal exertion and influence not less than six thousand of his fellow-townsmen are estimated to owe their escape from death by yellow fever. As a citizen, he was a signer of the Declaration of Independence, he moulded the constitution and law of his State, and was influential in establishing the common-school system of the country. As a scholar, he was an author, not only in medicine, but also in literature, philosophy, morals, and politics. He was decorated by four sovereigns of Europe, and elected to learned societies at home and abroad. By his genius and worth he honored his city and country, and elevated his art in the esteem of his fellow-citizens and of the sister professions. What has been his influence upon the medical profession of his city to give it that element of superior culture for which it is justly honored to-day! Though New York to-day is second to none in earnest, progressive medical work, we may well do homage to Philadelphia, whose faculty of medicine have done so much to maintain a profession of letters. Her first medical school had Benjamin Franklin as president; its professors were Rush, Kuhn—who had studied with Linnæus, Shippen—the student of John Hunter, Hewson and Fothergill, and Morgan—a pupil of William Hunter. A mantle of culture from the mother-country had fallen upon the physicians of this home of Rush.

But New York points with pride to many scholarly physicians—to Colden and Bard, to Beck, Francis, Watson, and Draper. And New England holds dear the names of Warren, Bigelow, and Jackson. To William Tully, a profound American scholar and physician, too little known beyond New England, I digress to pay a tribute of personal gratitude and justly-merited praise. William Tully was born at Saybrook, Connecticut, in 1785, and died at Springfield, Massachusetts, in 1859. As a scholar, he was an accomplished botanist, linguist and philologist, and a prominent collaborator of Webster's Dictionary. As a writer, he first contributed, with Dr. Miner, the celebrated "Essays on Fever," which circulated widely, excited discussion and criticism, and led to the formation of a medical school of opinion having many adherents to-day. At the time of his death he had prepared for the press two volumes of "Materia Medica and Pharmacology." They are original works, replete with the results of extended and thorough personal observation and experiment as to the physiological and therapeutic action of remedials, especially of those indigenous in America, as veratrum and conium. He advocated with zeal and power the supreme influence of the ganglionic system in the development of symptoms of disease and as the source of indications for the agents with which to treat them. He was president of the Castleton Medical School, and for eleven years Professor of Materia Medica in Yale College. He was instrumental in founding the well-known " Retreat for the Insane," at Hartford; and in the celebration of its fiftieth year of usefulness, in 1873, he is justly styled " the most learned man in his special departments of any in New England, or perhaps in the United States. Prof. Silliman has pronounced him to have been " one of the most erudite and philosophical scholars in the medical profession."

The impetus given to medical progress by the pursuit of the technical studies of our profession—the application of the microscope to anatomy, physiology, chemistry, and pathology, the methods of physical diagnosis, the certainty of diagnosis and treatment in the various specialties—leave us little to fear for the proficiency and training of the rising physicians to render them safe and successful practitioners of their art. But I would enter a plea for a medical study not immediately essential to the honest pursuit of medicine as a livelihood—the History of Medicine; with all its details of errors, false theories and superstition in the past, it points us back to a noble lineage of heroic men, thoughtful and earnest characters, devoted and careful observers, worthy of our admiration and imitation. It will dissipate the imputation that medicine of past times was crude and loathsome, and increase the individual strength, dignity, and self-respect, as being members of an ancient and honorable guild. We have terms in our medical nomenclature to-day which take us back to the medicine of Alexandria, of Greece, and Rome, perpetuating the anatomy of Herophilus and Galen, the surgery of Celsus. How many useful remedies, how many well-described diseases, date their literature from the observing and studious Arabian school, the Saracen physicians and writers of Spain, who translated the ancients, and saved their art when Europe was immersed in the darkness of the middle ages! As with literature and art, so medicine had its period of *renaissance* in the Italian school of the fifteenth and seventeenth centuries. Its great names are immortalized so long as we study the human frame—

Vesalius, Eustachius, Fallopius, Fontana, Casserius, Vieussens, Pacchioni, Fabricius, Malpighi. What have been the contributions to surgery and medicine of Boerhaave, Hoffman, Stahl, and Haller; of Paré, Bichat, Sauvage, Portal; of Willis and Winslow, Harvey and Hunter, Sydenham and Huxham? But the study of medical history, the history of medical discovery and progress, would be rewarded by a greater and positive benefit to our profession and the world. "It is a melancholy fact," says Sir Richard Steele, "verified by every day's observation, that the experience of the past is lost upon individuals and nations." The established facts of medicine are too often ignored, or rediscovered by the ill-informed. A greater familiarity with its history, a knowledge of the origin, the source of that which is settled, is the best point of departure for further inquiry and discovery.

Hippocrates practised incision and drainage of the pleural cavity, with much success—a procedure long abandoned, but being now revived.

Galen gave cold drinks, with benefit, in fevers, and employed immersion; for centuries this was overlooked. Again, in the beginning of this century, 1805, Currie published a full and complete manual of their treatment by water, with details and proofs, which were conclusive, yet to-day we are being led again, by Ziemssen and Liebermeister, to a renewal of that which should not have been lost. De Haen methodically employed the thermometer during the great fevers at Breslau. It was used by John Hunter. Currie employed the self-registering thermometer of M. Six in the mouth and axilla. It was put on record nearly one hundred and fifty years ago, and but to-day is it being utilized. In the sixteenth century the Italians found the quarantine and a stringent health code the certain means of limiting pestilence and infection, yet how long have sanitary corporations had full recognition and efficient value! Barker and Cheyne, in 1821, detailed the pavilion plan for hospitals, with diagrams of the five isolated buildings of their fever-hospital. Sydenham advocated fresh air and cold water in the treatment of small-pox, and was confirmed by the great fire of Pultney, where a hundred and more sick of this malady were turned out, to lie on beds, under hedges and the arches of bridges, with an immunity from deaths before unknown. Yet scarcely ten years have we had hospitals enforcing these conditions. Centuries ago, hæmoptysis, pulmonary congestion, and bronchitis, were conceded predisposing causes of phthisis; the certain relation was recognized between squalor and privation, exposure to damp and cold, the inflammatory diseases of the respiratory apparatus, and the ultimate development of consumption. For a century this evident relation of cause and effect has been set aside by the theory of specific tuberculosis; but the truth comes finally to recognition on a permanent basis of pathology. In the sixteenth century Chalin de Vinaria recognized the fatal error of bleeding in the plague and in all diseases of depressed vitality. He asserted it was safe only with the plethoric, reserving it, as he humorously said, for the priests and the indolent, whom it were well to dispose of. Though now discarded, how long did the lancet reign supreme! In an age of superstition, amid the confusion of the dancing mania, Paracelsus, the learned quack, presented a classification of the epilepsy, chorea, and hysterical seizures, which characterized that moral epidemic, as rational as that of neurologists to-day. He distinguished the true from the sensuous and imaginative forms. Yet for centuries the victims of these functional disorders were burned as witches, or shunned by society. There have been great and wise physicians in all ages of medicine.

Dr. Benjamin Rush, in his quaint address on "The Vices and Virtues of Physicians," gives as their fifth and last vice "obstinacy in adhering to old and unsuccessful modes of practice in diseases which have yielded to new remedies." Witness the reluctance fifty years ago to see in physical exploration of the chest a sure and definite guide to diagnosis of its diseases. But failure of the average physician to progress is more often the result of limited knowledge. He follows the directions of the Faculty, or men whose teachings moulded him for his life-work. Asserted medical facts, if of value, should not go long untested. Within a few years, Dr. Budd, of London, whom Sir Thomas Watson has called "one of the most strenuous cultivators of our science," has claimed to control scarlet-fever contagion in homes and asylums, without need of isolation. If true, it is one of our greatest contributions to medicine of recent times, yet it remains untested and unemployed; is known to but few.

Much is said, at the present time, about the "self-limitation of disease," "reliance upon Nature," "regimen and diet," and of the "expectant plan" of treatment. The public sentiment, too often the outgrowth of the doubts and controversies of the doctors, intelligently recognizes *hygiene*, but is skeptical of *therapeutics*. An eminent physician is said to have relinquished practice because he was tired of guessing. But these imputations are unjust. Mythology tells us that Hygeia, the goddess of health, was the daughter of Æsculapius, the father of medicine. Modern hygiene is the offspring of the physician's devotion to the study of the nature and treatment of disease. Nor have our greatest therapeutists ever ignored the simple remedies of Nature. The Greeks placed their temples of the healing art in the most healthful places; purified the air by fire and odoriferous incense; had baths, thermal springs, and gymnasia, as adjuvants. Such also were the practices of Asclepiades and Celsus in Rome. We point with pride to the school of Salerno and its famous code of health. The first of Sydenham's remedies was cool air, the second diet. First, the physician studying disease, then the detection of its cause, and subsequently the measures for prevention and abatement—such is the history of hygiene. Had not Sir George Baker known the symptoms of lead-poison, he could never have traced the cause of Devonshire colic—infection by lead would have continued, insidiously destroying, through the channels of tinctured food and drink, and numberless unsuspected sources; and colic, paralysis, and dementia, would have claimed as victims those who now are safe and healthy, Had Jenner been less of a symptomatologist, less a student of disease, less a physician, he would scarcely have pursued investigations whose result has been a substantial relief and abolition of the greatest scourge of society. It was the medical knowledge of Howard which enabled him to transform the prison, poorhouse, and asylum, from centres of pestilence—foci of contagion, dotting civilized lands—to healthful refuges for the vicious, the sick, and destitute. It was the science as well as the zeal and humanity of Pinel which reformed the treatment of the insane. Hygiene is a monument to the progress of medicine.

I have alluded to the influence which greater personal culture and literary talent will have to elevate the profession in the esteem of all the better classes. I believe that, with such better appreciation of our body by our clientage, there will arise a closer relation between us ; that the doctor of the future is destined to become the adviser and guide in much that pertains to safety and health, rather than an exclusive agency of necessity in sickness. If we define disease, with Reynolds, as "any condition of the organism which limits life in either its powers, enjoyment, or duration," surely our field, when confidence is stronger, will be broad. Abernethy used to advise his patients to peruse his works, as well as to take his medicine.

Have we not influence to avert the dangers of overwork, mental and physical; to dissuade from dissipation, to correct and prescribe the diet and drink that shall banish dyspepsias ; by suggestions and admonitions can we not protect our client from the incipient or exciting causes of phthisis ? We may tell the public of sunlight, of ventilation, and exercise. But, what the Sanitary Board is to the public, the doctor should be to the individual. It will be for him to say what are the relations of mind and body, when shall the child be sent to school, what educational stress will his individual health, his temperament and inherited constitution permit, what career is he warranted in undertaking.

The study of Nature, sanitary regulations, and regimen, are sources of immunity from disease, and often guide us in our arti-

2

ficial means of cure. But hygiene and hygienics can never supplant the therapeutist; it will ever hold true, "*Ars medica est id, quod est propter therapeuticen*"—everything in medicine is related to therapeutics. Let the present sources of infection and contamination be swept away, annihilating the loathsome plagues—products of miasm and effluvia. The mind will still become alienated, the brain overtaxed, the lungs inflamed, the heart enfeebled with age, and palpitating and faint in suffering and weakness; man will ever look to the physician, to one of greater knowledge and skill—not only as his counselor, guide, and friend, but chiefly to afford an earlier and more certain deliverance, or hasten the processes for which Nature points the way, but fails to accomplish if unaided. There is no greater fallacy in medicine than the assertion, "When Nature cannot work, the effect of art is void." We can well say of our art, as Laplace of science, "Our knowledge is trifling, our ignorance immense;" yet the doctor is a positive agency for good. Medicine is not the less a science because incomplete, imperfect, and unsettled. Is astronomy less a science, because there are undiscovered worlds and laws unexplained? Is mental science the less profound, though such a veil of mystery hangs over the subtile laws of mind. If we go back of the fifteenth century, before the brighter periods of medicine which Rénand terms the " age of renovation," even when medicine was ruled by theories and dogmas—mated with all these fallacies—were solid results, discoveries, truths slowly established but cumulative.

The profession, catholic and comprehensive in its principles, year by year discovers some new truth of therapeutics, and accepts every contribution of experience. Does the world owe nothing of its progress in civilization, its present safety, health, and happiness, to the healing art? Rome is said to owe her decline to the pestilences that devastated her armies, decimated her male population, and left her a prey to the barbarians. Small-pox was a serious obstacle to the progress of civilization in Europe. During one hundred years previous to the discovery of vaccination, forty million are estimated to have perished by its ravages. Do our merchants, our commerce, proceeding undisturbed by scurvy and cholera and yellow fever, owe nothing to medical science? In 1740, of a thousand seamen going to sea, six hundred and twenty-six never returned, destroyed by the scurvy. This disease crippled the marine of every nation. Yet by a therapeutic antidote—lemon-juice and the enforcement of partial vegetable diet at sea—it quickly disappeared; and in 1772 Captain Cook was enabled to sail round the world for three years, with but a single death. The *cure of disease* is the province of the physician. "*Felix qui potuit rerum cognoscere causas.*" But more important in daily life than theories of disease or knowledge of their cause is the certain knowledge of their means of cure. Cullen ascribed intermittent fever to the spasm of the skin, affected by miasm; Broussais to mucous phlegmasiæ; Brown to diffused excitability of the whole nervous system; to day we hold to a belief in the cryptogamic invasion of the blood; but, though we cannot discern the subtile nature of the disease, we can surely cure, by quinine and its analogues. Our recognition of pyæmia and septicæmia, the whole antiseptic method of prophylaxis and treatment, our certain means of assuaging pain, our resources for shortening and resolving disease—are they not such as to afford reason for our therapeutic faith, a faith which the sincere and honest physician must always have? " You cannot make a fire burn well if you put the wood on the andirons with a feeling of indifference." Says Macaulay: " We are quite sure that the improvement of medicine has far more than kept pace with the increase of disease, during the last three centuries. This is proved by the best possible evidence. The term of human life is decidedly longer in England than in any former age, respecting which we possess any information on which we can rely " (Southey's Colloquies,1830). How changed the mortality of great cities! that of London is said to be 50 per cent. less than it was two hundred years ago, and the average longevity of an Englishman is increased by several years. The history of medicine in the past and its rapid advances in the present justify us in designating it a science, imbued with the spirit of progress; and, after reviewing the steady and permanent contributions to its therapeutic resources in every age and country, we may indeed say of our art:

> " Step by step, since time began,
> I see the steady gain of man,
> That all of good the past hath had
> Remains, to make our own time glad."